Ernst Hartert

On the birds of the islands of Aruba, Curaçao and Bonaire

Ernst Hartert

On the birds of the islands of Aruba, Curaçao and Bonaire

ISBN/EAN: 9783337225018

Printed in Europe, USA, Canada, Australia, Japan

Cover: Foto ©berggeist007 / pixelio.de

More available books at **www.hansebooks.com**

[*From* THE IBIS *for July* 1893.]

On the Birds of the Islands of Aruba, Curaçao, and Bonaire.
By ERNST HARTERT.

(Plates VIII., IX.)

CONTENTS.

I. Introduction.

THE three Dutch West-Indian Islands of Aruba, Curaçao, and Bonaire (see map, Plate VIII.) are situated off the western part of the north coast of Venezuela. Aruba, being only about 16 miles from Cape San Roman, the north point of the peninsula of Paraguana, is nearest to the continent; Curaçao lies about 43 miles to the east; and Bonaire, the most oceanic of the three, still further to the east. Although near to the mainland, these islands do not, like Trinidad, belong geologically to the continent, but are of a different formation. They are surrounded by a coral-limestone belt, and for the most part are covered with a thick coral-limestone capping, and parts of coral-reefs are seen near Willemstad

[1]

on Curaçao, and on the east coast of Bonaire; while Aruba, for almost its entire length on the leeward side, is skirted by a coral-reef, inside of which is a calm and beautiful lagoon.

The interior parts of the islands consist of sedimentary rocks, in several places pierced by volcanic rocks, while on Aruba grey granite is said to predominate, and many quartz veins are found, containing a considerable amount of gold. Deposits of phosphate are distributed over the islands (*cf.* Blackburn, 'Aruba-Phosphate,' p. 5).

The geologist, Professor K. Martin, who explored the islands in 1885, came to the conclusion that they were formerly atolls; but his conclusion is questionable, for a similar coralline belt is found on many West-Indian islands.

The same naturalist (*cf.* 'Bericht über eine Reise nach Niederländisch Westindien,' 1887) came to the conclusion "that the islands of Aruba and Curaçao (the materials collected on Bonaire were too incomplete for any conclusions) are zoologically closely allied to the continent of South America, and, on the other hand, that the fauna of both differs in many points." Both these conclusions of the learned author—whose excellent book was of much service to me—must, however, be qualified in some way, although they are not altogether wrong. It is true that the greater part of the fauna is similar to that of the northern parts of Venezuela, but there are likewise a great many forms of West-Indian origin, and this not only among the birds, but also among the reptiles, and, according to Dr. Kobelt, very strikingly among the land-shells. The ornis and the whole fauna of the three islands are generally similar, although there are some remarkable differences.

The idea that the fauna of these islands is the same as that of the adjacent parts of the continent, together with their barren and rocky appearance from the sea, and the exaggerated reports of their heat and dryness, are perhaps the reasons why the ornis of Aruba and Bonaire remained unexplored until my researches, and why that of Curaçao has only quite recently, and incompletely, been explored.

No tropical forest is found on the islands, but trees of

[2]

different kinds abound, many of them introduced. The date-palm and the tamarind have been introduced and grow splendidly; the cocoanut-palm grows wherever it is planted. The bitter orange is grown in several large gardens to supply the valuable peel with which the famous Curaçao-liqueur is made. A great part of the islands is planted with the dividivi-tree (*Libidibi coriacea*), the husks of which are largely exported. The most characteristic features of the landscape, however, are the gigantic species of *Cereus, Opuntia,* and *Melocactus,* and the large fields of *Aloë.* The largest tree I saw on the islands was an old and fine *Eriodendron,* at the foot of Mt. Christoffel, not far from Savonet. On all the islands the *Rhizophora* grows here and there on the coast, and in many places over a great extent.

As I have stated above, the accounts of the extraordinary dryness of these islands are exaggerated. The year 1892 was, it is true, an unusually wet one, and 1885, the year in which Professor Martin visited the islands, was perhaps one of the driest of the century. Having read the description of Martin and that of Herr Peters (J. f. O. 1892, p. 105) in manuscript, I did not expect to find much vegetation. Great, therefore, were my joy and astonishment when on the 3rd of June, at daybreak, I saw the picturesque rocks of Curaçao before me, sparsely but thoroughly covered with the freshest green.

That day I could not leave the steamer for hours, and the shops of Willemstad were not opened before 11 o'clock, on account of the pouring rain—and rain troubled me more than once after this on these "rainless" islands. The vegetation, therefore, was rather rich during my stay, and many lovely flowers were seen, especially on the slopes of the Christoffel, where I found three species of orchids. These plants, of course, must be indigenous, and trees of several metres in girth and of considerable height cannot grow up and disappear at short intervals.

Fresh water is very scarce and valuable at times, and there are only one or two places on each island with natural springs; but there are beds of rivulets on the slopes of the

Christoffel, and also on Aruba and Bonaire, which must sometimes be filled with water.

The spring at Hato on Curaçao is the only known habitat of a little fish (*Pœcilia vandepolli*). Of water-beetles I caught examples of several species on Curaçao and Bonaire. Mammals are very scarce; I found only one species of Bat, and the European Rat and Common Mouse. A species of Hare is common on Aruba and Curaçao, but is not found on Bonaire. The Venezuelan Deer has been introduced in Curaçao, and a great number of goats run all over the islands, and, no doubt, do much harm to the vegetation.

Bird-life is abundant, and there are many species that could not be more numerous arywhere. Breeding-places of sea-birds are found only on Aruba and Bonaire, and they are not very extensive, but enormous numbers of sea-birds breed on " Los Aves," east of Bonaire, and " The Monks," west of Aruba. Now and then boats go there and bring large quantities of eggs to Curaçao for sale as food. Unfortunately I was too late, so I was told, for the egg-season, and therefore I did not visit those uninhabited rocks, thinking that the results, at that time of the year, would not be sufficient to repay the costs and hardships of such a trip in one of the fishing-boats, but I should advise future explorers to go there at the proper time.

The trade-wind blows over the islands almost incessantly, with more or less vigour, and on exposed parts all the trees lie over to the westward, presenting a peculiarly stormy appearance. ·The strong wind is, perhaps, one of the chief reasons why insect-life is so scarce. Reptiles, however, are very numerous, but not many species occur. Bird-life, too, is influenced by the wind, for birds are more numerous in places where the trade-wind does not penetrate, while on exposed plateaus they are very scarce as a rule.

On Curaçao I collected for three weeks altogether, and visited many places in all parts of the island, staying 10 days at Savonet at the foot of the Christoffel, ascending to the top of this mountain, staying at Willemstad, Beekenburg, and Hato, and exploring the country round these places.

[4]

On Aruba I remained 16 days, and about the same time on Bonaire. I visited many places on both these islands, the hills and the plains, the barest and driest places, and the richest and best-wooded parts, and, with the help of my wife, collected birds vigorously all the time.

The almost continuous sunshine, the beautiful clear atmosphere, the salubrious and wonderfully warm temperature, never or seldom rising to an unendurable heat, and the picturesque scenery gave me pleasures which can never be forgotten.

I wish here to express my sincerest thanks to my friend Freiherr Hans von Berlepsch, in whose museum and company I compared and studied the greater part of my collection on my return.

The types and the first pick of all my skins are in Mr. Walter Rothschild's Museum at Tring, and most of the duplicates, containing some co-types, are in Freiherr von Berlepsch's collection.

II. *Birds of Aruba.*

Aruba is the driest and in most parts the barest of the three islands. There are several good breeding-places for sea-birds. I was on the island from the 21st of June to the 5th of July, the season when but few wanderers from the north can be expected, and therefore most of the birds that I collected are residents. The island is, of course, resorted to by several northern birds in winter, and Venezuelan birds are said to visit it often in autumn.

I am much obliged to several inhabitants of Aruba, and above all to our amiable host, Dr. Coates Cole, the English surgeon of the island.

Nothing has as yet been written on the birds of Aruba. But Prof. Martin mentions in his book that he saw a *Conurus* that was different from *C. pertinax* of Curaçao, a large kind of parrot, a *Mimus*, two Humming-birds, an *Icterus*, an *Ortyx*, and a *Pelecanus.* Besides this, Mr. G. N. Lawrence has described a living parrot from Aruba under the name of *Chrysotis canifrons*, and so long ago as 1658

[5]

the French traveller Rochefort stated that he saw two kinds
of Humming-birds, of which one was the smallest and the
most beautiful he knew, on Aruba (Berl. J. f. O. 1892, p. 65).

All the above-mentioned birds were found by us, and
collected in sufficient numbers to enable us to identify them.

1. MIMUS GILVUS ROSTRATUS, Ridgw. Proc. U. S. N. M.
1884, p. 173 (Curaçao); Berl. J. f. O. 1892, p. 74 (Curaçao);
Peters, J. f. O. 1892, p. 114.

(1) ♂ ad. sect.   Aruba, 22 vi. 1892.   Wing 4·4 inches,
tail 4·6, culmen 0·9, tarsus 1·4.

(2) ♂ ad. sect.   Aruba, 23 vi. 92.   Wing 4·25 inches,
tail 4·5, culmen 0·9, tarsus 1·4.

The specimens of this bird from Aruba agree in every
respect with those from Curaçao.   Berlepsch (*l. c.*) has said
much about this form, which to a certain extent varies indi-
vidually.   It certainly does not deserve more than subspecific
rank.

The "*Tjutjubi*" is not rare on Aruba, but less numerous
than on Curaçao.

The iris is dark orange-brown, bill and feet black.

Its food consists of fruits, chiefly that of the *Cereus*, and
beetles.

The nest is a large and somewhat loose structure, mostly
placed in the dividivi-trees.   The eggs are four or five in
number, with the well-known coloration of those of the other
forms of *Mimus*, all much of the *Turdus*-type, thereby con-
firming my opinion that *Mimus* should not be removed too far
from the Thrushes.

The name "*Tjutjubi*" is taken from an often-heard note
of this bird, closely resembling these syllables.

I found fresh eggs on Curaçao in the middle of June, and
hard-set ones at the beginning of August.   In the meantime
I frequently met with quite young birds flying about, and
also found some nestlings.

The "*Tjutjubi*," sitting on the top of the high *Cereus*, and
often singing its pleasant notes even from the roofs of the
houses, is one of the most characteristic features of the avifauna of Curaçao, Bonaire, and Aruba.

[6]

2. Dendrœca rufopileata, Ridgw.

Dendroica rufopileata, Ridgw. Proc. U. S. N. M. 1884, p. 173 (Curaçao) ; Berl. J. f. O. 1892, p. 76 (Curaçao).

This bird is very common on Curaçao and Bonaire, but rather scarce on Aruba, where I found it in a few localities only, and in small numbers. I have only three skins from Aruba. For further details see below, p. 311.

3. Certhiola uropygialis (Berl.).

Cœreba uropygialis, Berl. J. f. O. 1892, p. 77 (Curaçao).

Not rare on Curaçao and Bonaire, but much less numerous on Aruba than on the other two islands. Aruban specimens agree in every respect with those from Curaçao.

4. Euetheia sharpei, Hartert, Bull. B. O. C. vii. p. xxxvii. Not rare on Aruba.

5. Zonotrichia pileata (Bodd.), Berl. J. f. O. 1892, p. 82 (Curaçao).

This bird is common on Curaçao, where it is met with everywhere. On Aruba it is very rare, and unknown on Bonaire. The single specimen I have from Aruba has a very stout bill, but otherwise agrees entirely with those from Curaçao.

6. Icterus xanthornus curaçaoensis (Ridgw.), Berl. J. f. O. 1892, p. 82 (Curaçao).

Icterus curaçaoensis, Ridgw. Proc. U. S. N. M. 1884, p. 174; Scl. Cat. B. B. M. xi. p. 381 (1886).

Berlepsch has carefully compared several specimens of this bird from Curaçao with those from other localities, and has pointed out that there is nothing to distinguish the Curaçao form but its longer bill. It is true that the bills of the Curaçao birds are longer than usual, and the colour is also a little paler as a rule; it may therefore stand as a subspecific form of *Icterus xanthornus.*

The specimens from Curaçao all agree, but two males from Aruba have the bills shorter and stronger, and also the yellow colour brighter and more tinged with orange. They therefore point more to the true continental *Icterus xan-*

[7]

thornus. This is another reason for considering the Curaçao bird merely a subspecies. The measurements of my specimens are as follows :—

♂ ad. sect.	Aruba, 23 vi.	Culmen 0·9 inch, wing 3·9.
♂ ad. sect.	Aruba, 1 vii.	Culmen 0·88 inch, wing 3·8.
♂ sect.	Curaçao, 8 vi.	Culmen 1 inch, wing 3·7.
♀ sect.	Curaçao, 13 vi.	Culmen 1·05 inch, wing 3·5.
♀ sect.	Curaçao, 14 vi.	Culmen 1·06 inch, wing 3·7.
♀ sect.	Curaçao, 16 vi.	Culmen 1 inch, wing 3·5.
♂ sect.	Curaçao, 2 viii.	Culmen 1·05 inch, wing 3·7.

It is, I think, very interesting that the birds from Aruba, the island nearest to the continent, agree better with the continental form than those from Curaçao. The bird is equally common on all three islands, but only where it finds sufficient trees in which to build its long hanging nest. I have not procured skins from Bonaire, but the birds there agree with those from Curaçao. I got an egg on the 22nd of July on Bonaire. The colour is of a pale bluish white, sparingly covered with long and fine deeper lying cinereous hair-lines and overlaid patches and lines, like Arabian letters, of a deep purplish brown, more frequent on the thicker end. It measures 0·93 × 0·67 inch, and the weight of it is 250 milligrammes.

The bird is sometimes kept in captivity, but is not much appreciated. Its piping notes are less clear than those of *Icterus vulgaris,* and they produce many screeching and mewing sounds. Herr Peters (J. f. O. 1892, p. 114) thinks that the Curaçao form has a different note from that of the continental *I. xanthornus,* but this seems to be imagination. I have heard the *I. xanthornus,* and both, without doubt, " speak the same language."

In the " papiamento "—the mixed dialect of Spanish, Portuguese, and Dutch spoken on these islands—this bird is called " *Trupial cacho,*" *i. e.* Dog-Trupial. On Aruba it is called " *Gonzalito.*"

The sexes in the adult bird are alike, but young birds have greenish-olive tails. They seem to retain the immature plumage for some years, as has been stated by Baird (B. N.

[8]

Amer. p. 540) to be the case with other species of *Icterus*. I dissected specimens with green tails that had swollen testes and had paired. The black tail is assumed by changing the colour, not by moult, as two of my specimens clearly show.

7. ICTERUS ICTERUS (Linn.).

Oriolus icterus, Linn. Syst. Nat. i. p. 161 (1766).
Icterus vulgaris, Daud. Tr. d'Orn. ii. p. 430; Scl. Cat. B. B. M. xi. p. 382.
Icterus vulgaris (subsp.?), Peters, J. f. O. 1892, p. 114 (Curaçao).

Peters (*l. c.*) says that this bird occurs on Curaçao, and is said to be paler than the continental form. After carefully comparing my specimens with those in the British Museum, I must say that my birds, on the contrary, have very bright colours, and that they are indistinguishable from the continental *Icterus icterus*. My specimens are rather short-winged, but those from Santa Marta in the British Museum are quite similar. An example from Carupano is a good deal larger, but there are others intermediate. A specimen from an unknown locality in H. v. Berlepsch's museum has white spots on the outer rectrices, and one from Carthagena is rather paler than my birds.

♂ ad. sect. Aruba, 27 vi. Total length about 9 inches, wing 4·4, tail 4, tarsus 1·2, culmen 1·28.

♀ ad. sect. Aruba, 27 vi. Total length about 9 inches, wing 4·3, tail 3·9, tarsus 1·2, culmen 1·3.

♂ sect. Curaçao, 2 viii. Total length about 9·5 inches, wing 4·35, tail 4, tarsus 1·3, culmen 1·37.

♀ sect. Curaçao, 2 viii. Total length about 8·75 inches, wing 4·15, tail 3·9, tarsus 1·25, culmen 1·24.

My specimens are in a somewhat worn plumage. I did not find any nests; but, as everybody on Curaçao knows, they are totally different from those of *Icterus xanthornus* in not having the long tube.

This bird is much appreciated as a cage-bird on account of its pure flute-like notes, and is often sent for sale from Venezuela.

[9]

This species is not rare in certain places, such as the rocky hills covered with brushwood and cactus, both on Aruba and Curaçao, but it is absent from Bonaire, thus indicating its immigration from the continent. I saw it in the bush on St. Thomas, where it has already been stated to occur by Ridgway. It may have been introduced into that island; but, on account of other affinities between the ornis of St. Thomas and that of Curaçao, this is very questionable.

8. MYIARCHUS BREVIPENNIS, Hartert, Bull. B. O. C. iii. p. xii; id. Ibis, 1893, p. 123.

I have compared this new insular form of *Myiarchus* with specimens in Berlepsch's museum and with the fine series in the British Museum, and find that it is closely allied to *Myiarchus tyrannulus* (*cf.* Scl. Cat. B. B. M. xiv. p. 251), but readily distinguishable from it by its shorter wings and tail, longer tarsus, the more olive-greyish and less brownish colour of the upper parts, and the blackish lower mandible, which in *M. tyrannulus* is pale brown.

It is remarkable that in Venezuela the true *M. tyrannulus* occurs, and that the Island of Grenada is inhabited by another species, *M. oberi*, Lawr. Sclater (*l. c.*) has united *M. oberi* with *M. tyrannulus*; but the specimens now in the British Museum and in Berlepsch's collection show that *M. oberi* is a very distinct species. It differs in the much greater extent of the rusty colour on the inner webs of the rectrices, decidedly darker upper surface, longer bill, and longer wings and tail, thus pointing more to *M. mexicanus* in its size, but not in the colour of the back. Specimens from the three islands Aruba, Curaçao, and Bonaire are quite similar. Total length about 7·3 inches, wing 3·4 to 3·59, tail 3·3 to 3·5, culmen 0·7 to 0·8, tarsus 0·75 to 0·85.

9. SUBLEGATUS GLABER, Scl. & Salv. P. Z. S. 1868, p. 171, pl. xiii. (Caracas); Berl. J. f. O. 1892, p. 84 (Curaçao).
Not very rare.

10. TYRANNUS DOMINICENSIS (Gm.).
This bird seems to be very rare on Aruba, where I did not obtain it, but once saw a pair. That this species is rarest on

Aruba and more common on the other islands seems to point to the fact that it is a West-Indian form.

11. CHRYSOLAMPIS MOSQUITUS (Linn.).

Common on flowering trees. While on Curaçao in the beginning of June these birds were in moult, and it was impossible to obtain males in good plumage: they began to get out of their moult by the end of my stay on Aruba.

When I met with this beautiful Humming-bird I did not know there was any question to settle about it, and did not pay especial attention to it. I did, however, collect a series of fine adult males, and, chiefly owing to the efforts of my wife, eight specimens in dull plumage, all well skinned and dissected. In looking over Mr. Salvin's description in the Cat. B. xvi. p. 114, I find the adult female described as having the lateral tail-feathers bronzy black, but my skins contravene this statement. According to my series *the adult female has the rectrices chestnut-red, with a broad subterminal band of a purplish steel-blue, and tipped with white.* They appear to have sometimes, if very aged, some glittering feathers along the middle of the throat. The young of both sexes—according to my collection—have the tail purplish black, and I have (in my own collection and among a number of trade-skins) many intermediate stages. Gould and Lesson have both figured the females as they really are, with the red tail. As regards the name, it should be written *mosquitus* and not *moschitus,* as shown before by Berlepsch. Linnæus in his Syst. Nat. ed. x. p. 120 (1758), as well as in ed. xii. p. 192 (1766), wrote it *mosquitus,* and it was only changed to *moschitus* by Gmelin. Linnæus very probably meant to designate it a small mosquito-like bird.

12. CHLOROSTILBON CARIBÆUS, Lawr. Ann. Lyc. N. H. New York, x. (1874) p. 13.

Not rare, but rather less common than the foregoing species. Badly in moult. Berlepsch has shown (J. f. O. 1892, p. 87) that the name *C. atala* of Lesson is very doubtful, and that the acceptance of Lawrence's name is advisable.

[11]

13. CONURUS ARUBENSIS, Hartert, Bull. B. O. C. iv. p. xvi (1892).

Adult male and female. Forehead pale yellow for about 0·3 inch; top of the head distinctly tinged with blue; circle round the eyes very narrow above, broader below, pale orange-yellow; lores, cheeks, and sides of the head mixed light brown and very pale orange-yellow, the feathers, especially those on the ear-coverts, being yellowish in the middle and bordered with brown; throat and upper breast yellowish brown. Of the same size as *Conurus æruginosus*, but tail longer as a rule.

This form of *Conurus* is closely allied to *C. æruginosus* from Guiana and Venezuela, from which it differs only in the lighter colour of the forehead, sides of the head, and throat, and I believe also in a somewhat longer tail. As my four specimens from Aruba differ in these points from nine skins from British Guiana in Mus. W. Rothschild, from all the skins from different localities in the British Museum, and from skins from Venezuela in Mus. H. v. Berlepsch, I believe I am right in distinguishing it as a new island form.

♂ sect. Aruba, 22 vi. Iris straw-yellow; bill horn-brown; feet deep brown. Total length 9·7 inches, wing 5·3, tail 4·9, culmen 0·9, tarsus 0·5.

♂ sect. Aruba, 23 vi. Iris pale yellow. Total length about 9·6 inches, wing 5·3, tail 5·1, culmen 0·95, tarsus 0·5.

♀ ad. sect. Aruba, 2 vii. Wing 5·1 inches, tail 4·85, culmen 0·85.

♂ ad. sect. Aruba, 2 vii. Wing 5·45 inches, tail 4·9, culmen 0·78.

In fifteen specimens of *C. æruginosus* from Guiana and Venezuela the tail measures 4·25 to 4·6 inches, the wing 5·3 to 5·65, the culmen 0·85 on the average. In Salvadori's description (Cat. B. xx. p. 196) the length of the bill is given as 0·28, which is evidently a misprint for 0·78 or 0·88.

Conurus arubensis might, on account of its somewhat yellowish sides of the head, be looked upon as a form intermediate between the continental *C. æruginosus* and *C. pertinax* from Curaçao and St. Thomas. To those naturalists who unite these two forms, this statement may appear to be a rather bold one. But it is not wrong to say that those who cannot distinguish between *C. pertinax* and *C. æruginosus* are not well acquainted with these birds. I myself did not know them when, two years ago, in my 'Katalog Vogels. Mus. Senckenberg.' (p. 156), I ventured to unite the two species, having been (like Finsch, Schlegel, and others) misled by young specimens of *C. pertinax* and by inexact localities, so that the distribution could not be studied. With more material at hand it might not be wrong to regard this new form as a subspecies of *C. æruginosus*, as it is close to it, and specimens might easily be found that very nearly approach it, but I prefer to keep it as a species, all the more on account of its isolated habitat.

C. arubensis is very common everywhere on Aruba. The first morning when out shooting with my friend Dr. Cole, I obtained a specimen of it. Thinking that it was the common continental form, I was content to pick up a specimen occasionally, and brought home four skins only. My much honoured friend, Count Tommaso Salvadori, first called my attention to the light-coloured foreheads and cheeks in my skins as soon as he saw them, and I was glad that I found the surmises of this great ornithologist well founded.

C. arubensis is similar in its habits and screaming voice to *C. pertinax*, and also lays its eggs in holes dug out in old ants' nests and trees, and in the natural caves and holes in the lime rocks. Its food consists mostly of the fruits of *Cereus, Melocactus,* and other plants.

14. CHRYSOTIS OCHROPTERA (Gm.).

Psittacus amazonicus gutture luteo, Briss. Orn. i. p. 287.

Le Perroquet à épaulettes jaunes, Levaill. Perr. pls. 98, 98 bis.

Chrysotis ochroptera, Reichen. Vogelb. pl. i. fig. 5 ; Salvad. Cat. B. B. M. xx. p. 288.

Chrysotis canifrons, Lawr. Ann. N. Y. Acad. Sci. ii. p. 381 (1883) (Isl. of Aruba); Salvad. Cat. B. B. M. xx. p. 272 (note).

This beautiful Amazon, of which, in spite of the numbers

[13]

that are kept in confinement, specimens procured in a wild state are so rare in museums that its habitat could only be given with a query in Salvadori's Catalogue of the Parrots (*l. c.*), inhabits the Island of Aruba. It might not be out of place here to state that it is also common in the lowland forests of the district of Coro, and in other parts of Venezuela, whence large numbers are sent to the bird-shops of the larger towns of Venezuela and to Curaçao.

I procured three adult males of this fine bird. They are very bright-coloured, forehead and lores white with a faint ashy hue, the greater part of the top of the head, and in all three specimens some of the feathers on the neck also, rich yellow with rosy-orange bases ; the entire sides of the head and chin of the same colour, corresponding with Brisson's description and Levaillant's very good figure. The whole of the cubital edge, the bend of the wing, and nearly all of the lesser wing-coverts bright yellow ("épaulettes jaunes" of Levaillant); thighs bright yellow with a rosy tinge at the bases of the feathers; bill whitish horn-colour; iris orange-red, shading into orange-yellow inwards; feet dark grey. My specimens are coloured thus, but in captivity these birds often besmear the forehead with dirt, as many also do in a wild state with the sticky juice of the fruits of *Cactus*. In European Museums, where all or nearly all the specimens are from individuals that have died in confinement, the yellow is often not so much extended. The plumage of the perfectly adult bird may perhaps never be acquired in Europe, where most specimens of this Parrot are brought when very young. In the immature bird the yellow on the head is less diffused round the eyes; the chin and cheeks are pale bluish, and probably quite blue in birds lately from the nest, the bluish colour getting more and more mixed and overspread with yellow as the bird gets older (as I observed in my live specimens from Coro that I brought home with me); bend of the wing greenish, and cubital edge not so bright yellow; thighs pale yellow ; iris reddish brown.

The cubital edge is always yellow, except in its innermost corner, where occasionally a few red feathers appear. Some-

[14]

times some whitish feathers can be seen on the chin. Total length about 13·5 inches, wing 8·4 to 8·7, tail 5·3, culmen 1·4 to 1·45, height of upper mandible 0·65.

This Parrot is not rare in the more wooded and rocky parts of the island, but is somewhat shy and not easily to be obtained in numbers. It is said to breed in hollow trees. A live specimen from Coro in Venezuela in the possession of Dr. Cole was in every respect like my collected specimens, but in Europe such finely coloured birds are very seldom to be seen alive.

There can be no doubt that *Chrysotis canifrons* of Lawrence (*l. c.*) was described from an example of this species with a dirty forehead, such as I have seen in several cases. It was based on a living specimen brought to New York from Aruba, but the type has been lost. Among my specimens of *Chrysotis rothschildi* from Bonaire are several that show a somewhat ashy colour on the forehead.

15. POLYBORUS CHERIWAY (Jacq.).

Not rare on all the three islands. I have a skin from the peninsula of Paraguana, Venezuela, collected by Herr Ludwig. It is similar to one shot for me by Dr. Cole on Aruba, which I did not skin. The skin from Paraguana agrees with those from other countries. On Bonaire this bird places its nest on trees.

Local name "*Warawara.*"

16. TINNUNCULUS SPARVERIUS BREVIPENNIS, Berl. J. f. O. 1892, p. 91 (Curaçao).

The "*Kinikini*" is not rare on all the three islands. I have several specimens of both sexes, and find Berlepsch's characters constant. I agree with him in distinguishing it subspecifically—the difficulties of such forms as those of *Tinnunculus sparverius* being best met by dividing them into several subspecies. The wings of the males measure 6·5 to 6·8 inches, tails 4·9 to 5, tarsi 1·4; the wings of the females 6·5 to 6·7, tails 4·8 to 5, tarsi 1·3 to 1·4.

The rufous spotting of the crown varies much, and is usually almost or quite absent.

17. BUTEO ALBICAUDATUS COLONUS, Berl. J. f. O. 1892, pp. 89 & 91 (Curaçao).

Seen, but not procured. Possibly, however, it was not this form, but the continental one.

18. ZENAIDA VINACEO-RUFA, Ridgw. Proc. U. S. Nat. Mus. 1884, p. 176 (Curaçao).

Very common. Identical with examples from Curaçao.

19. COLUMBA GYMNOPHTHALMA, Temm.

Not rare on Aruba. Identical with specimens from Curaçao.

20. COLUMBIGALLINA PASSERINA PERPALLIDA, subsp. nov.

Columbigallina passerina, Berl. J. f. O. 1892, p. 97 (Curaçao).

There is hardly a bird that presents more local variation than this pretty little Ground-Dove. North-American authors distinguish between the form inhabiting the "South-Atlantic and Gulf States" and the one reaching from the South-western States throughout Mexico. The former they used to call *C. passerina,* but later on it was named *C. passerina purpurea* by Maynard, and has quite recently been renamed *C. passerina terrestris* by Chapman.

I have collected a large series from the three Dutch West-Indian Islands. All of them differ from the forms of other countries, that I have seen, in their pale colour throughout, and especially on the under tail-coverts, in the rather shorter wing, and in the base of the bill being *yellow,* not *red.* On Porto Rico and St. Thomas I shot examples of a different race, much richer and darker in colour everywhere; the base of the bill deep red, not yellow, and the wing also short.

The typical form of Linnæus's *C. passerina* must be the Jamaica bird (*cf.* Berl. J. f. O. 1892, p. 97), which, if anything, has the wing a little longer on the average than the one from Porto Rico, and stands, as regards coloration, between the pale and dark forms. The Eastern North-American bird is closely allied to this form, but it is said to have the base of the bill red, and the wing is certainly a little longer. It might therefore be distinguished subspecifically, *more Americanorum.* The Mexican pale form, however, is quite distinct, and the rich-coloured birds from Porto Rico deserve attention. The latter correspond with Ridgway's description of *C. passerina socorrensis,* but are probably distinct from it. The form from Grand Cayman described as *C. passerina insularis* is probably the same as the true typical Jamaican *C. passerina.* My pale birds from Curaçao are in colour nearest to the Mexican bird, but the wings are a little shorter, the colour still a trifle paler, and the base of the bill yellow, instead of red, as it is said to be in *C. passerina pallescens* from Mexico.

All the South-American examples of *C. passerina* seem to be very closely allied to the true Jamaican *C. passerina,* although extremely variable. It is of course safer, to avoid mistakes, to unite all the forms together, but I am not prepared to do this. I have not seen Barbadian specimens, on which Bonaparte's name *C. trochila* (Consp. Avium, ii. p. 6) was based.

The Ground-Dove of Curaçao and its sister islands has the bill "deep brown at the tip, the basal portion pale orange-yellow or pale yellow, near the nostrils light yellow. Iris lilac and red. Naked ring round the eye light yellow. Feet light flesh-colour." Wing 3·05 to 3·15 inches.

This Dove is extremely common on all the three islands, and is known as "*Tortolica.*" The nest is placed in bushes and trees, but mostly on the prickly branches of the *Opuntia* or *Cereus.* It contains two eggs. I found two broods, and I was told that some of them breed in every month of the year. The eggs are of an elliptical ovate and elliptical oval form, varying into elongate ovate, occasionally ovate, and even nearly fusiform; they measure 19·6 × 15, 20 × 17, 21 × 17 to 22 × 16, 22 × 17, 23 × 17, 23·5 × 15·7, 23·6 × 16·5 to 23 × 17·5 mm.

21. LEPTOPTILA VERREAUXI, Bp.

Appears to be very rare. I saw it only once in Aruba.

22. EUPSYCHORTYX CRISTATUS (L.).

Eupsychortyx gouldi, Berl. J. f. O. 1892, p. 100 in the text (Curaçao).

[17]

Linné's description of *Tetrao cristatus* is founded on
Brisson's *Coturnix mexicana cristata* (Orn. i. p. 260, pl. xxv.).
Brisson is of course wrong in his locality, but clearly figures
and describes the form from Curaçao. As this island was
always visited by ships, the Abbé Aubry's Museum had very
likely got specimens from there. Gould (Monogr. Odontoph.
p. 16, pl. ix.) figures the present bird as *E. cristatus*, but
his localities are no doubt partly, if not altogether, wrong.
Berlepsch (*l. c.*) has already well described his *E. gouldi*, and
has pointed out in what respects it differs from *E. sonninii*.
The most obvious are the black stripes above and below the
ear-coverts, which never show in the species from Guiana, and
the colour of the underparts. The species from Colombia
(Bogotá) is, beneath, more similar to *E. cristatus*, while its
head is more like that of *E. sonninii*. There can be no
doubt that all three species are quite distinct.

Berlepsch had received only one skin—a female, as stated
by Peters, but in the fine plumage of the adult male, as
figured by Gould. My series contains but one female, and
this is similar to the males in plumage, but has the ear-
coverts brown and merely traces of the black stripes on the
sides of the head. I believe that the female gets the black
stripes when fully adult, and that Peters's statement was
right, while Gould has figured young birds as females, for I
have three young specimens in different stages—one male,
one female, and one with the sex not determined. All these
three are alike and agree with Gould's figures of the so-
called females. The wings of my adult males measure from
3·9 to 4·1 inches, tarsus 1. The iris is dark brown, bill black,
and feet brownish grey.

This pretty bird is not rare on Aruba and Curaçao, but is
not found everywhere. The natives call it " *Socklé*," a name
derived from its note, which is uttered very frequently. It
is much esteemed as food, and sometimes sold in the market
alive.

This bird is not easy to obtain in any great numbers
without a dog, as it does not care to fly and is difficult to be
seen in grassy places. It is not found on Bonaire.

[18]

I am quite sure that Gould's habitat ("Mexico") for this species is wrong, for recent explorers have not found it there; but I have reason to believe that the bird occurs in Venezuela, where *E. sonninii* is also found, but probably not in the same localities.

23. ARDEA TRICOLOR, Müll.

Not plentiful, but of regular occurrence on Aruba and Bonaire. Identical with South-American specimens, but different from the Mexican subspecies, which is spread over the West Indies. Culmen 3·65 inches, wing 9·4, tarsus 3·5.

24. ARDEA CANDIDISSIMA, Gm.

Seen on Aruba and on Bonaire. Bill—posterior portion bluish flesh-colour, anterior half blackish horn-colour. Iris silvery white. Legs sky-blue, large scales in front of tarsus black. Total length 27 inches, wing 12, culmen 3·5.

25. BUTORIDES VIRESCENS (Linn.).

Found on all the three islands. My specimens agree with other examples of this species, but the wings are rather shorter, measuring only 6·5 inches; culmen 2·25, tail 2·4, tarsus 1·8.

This is probably the bird called *B. striata* by Peters (J. f. O. 1892, p. 121).

26. PHŒNICOPTERUS sp. inc.

A Flamingo was seen and shot at by Dr. Cole. It is said to be rare, and a straggler only.

27. CHARADRIUS SQUATAROLA, Linn.

I saw a few of this species and shot a male on the 24th June on Aruba.

28. STREPSILAS INTERPRES, Linn.

I saw three individuals and shot one on the 2nd July on Aruba.

29. ÆGIALITIS RUFINUCHA (Ridgw.).

Ægialitis wilsonius, var. *rufinuchus*, Ridgw. Am. Nat. viii. 1874, p. 109; id. Man. N. Am. B. p. 175. (*Hab.* West Indies and Atlantic coast of S. America to Bahia.)

z 2

This Plover is common, and undoubtedly breeds, on Aruba and Bonaire. I think it belongs to Ridgway's subspecies, but that it deserves specific rank. Two adult males in very fine plumage have no traces of a black band across the chest. Lores decidedly rusty. Culmen 0·83 to 0·85 inch, wing 4·45 to 4·5, tarsus 1·1.

The black band across the chest is probably always replaced in the adult male by a rusty rufous band.

30. HÆMATOPUS PALLIATUS (Temm.).

I only once saw a specimen of this Oyster-catcher on the reef on Aruba and fired at it, but unfortunately missed it.

31. TOTANUS FLAVIPES (Gm.).

This bird was common on Aruba on the 22nd June, when Dr. Cole shot two specimens.

32. PELECANUS FUSCUS, Linn.

Extremely common and not at all shy.

33. FREGATA AQUILA (Linn.).

Schlegel (Mus. d. Pays-Bas), Oates (B. Brit. Burm.), and others are of opinion that the white-breasted specimens of this species are young birds, but Ridgway (B. N. Amer. and Man. N. Am. B.) has already well described the plumage of the adult female as well as that of the young, which has the whole head white. My male example agrees perfectly with specimens from the Pacific and Madagascar*. The females have much larger bills than the males. My specimens measure :—

♂ ad. Aruba, 3 vii. Culmen 5·2 inches, wing 23·6, tail 17.

♀ ♀ ad. Aruba, 3 vii. Culmen 5·5 to 5·6 inches, wing 23 to 24·5, tail 15·5.

34. PHALACROCORAX BRASILIANUS (Gm.).

Great flocks of this Cormorant were seen on Aruba, but were very difficult to approach.

* Hartlaub ('Vögel Madagascars,' p. 399) mentions only *Fregata minor* from that island, but examples of both these very distinct species have been recently received by the Tring Museum from Madagascar.

35. STERNA MAXIMA, Bodd.

I have two specimens of this fine Tern, one from Aruba
and one from Bonaire, but the bills and wings seem to be
shorter than in *Sterna maxima* proper, and the comparative
measurements of the bill do not quite correspond with those
given by Ridgway in his 'Manual.' They do not, however,
belong to the Pacific *S. elegans*, nor to Saunders's Atlantic
S. eurygnatha. This specialist in Laridæ has seen one of my
specimens and admitted it to be *S. maxima*, but I think that
a larger series would be of interest, and might possibly lead
to the establishment of a South West-Indian subspecies of
S. maxima. The culmen in my specimens measures 2·25 to
2·36 inches, wing 13·3 to 13·6, tarsus 1·1. Iris brown,
bill orange, feet black. A male and female from Georgia
are similar in the form of the bill, but the latter is 0·3 inch
longer, and the wings measure 14·5 inches.

This Tern is not common on these islands and is somewhat shy, but I saw it several times on the coasts of Aruba,
Curaçao, and Bonaire.

36. STERNA HIRUNDO, Linn.

I have seen this Tern flying about along the coasts of
Aruba and on Bonaire, and I believe also on Curaçao, but I
have brought home only two skins. These agree with the
European *Sterna hirundo* (= *Sterna fluviatilis* of Naumann)
in appearance, but are much smaller and the bill somewhat
less pointed. In fresh specimens it seemed to me that the
abdomen was somewhat less greyish and of a more violet
tint. I also believe that the black cap does not reach quite
so low down on the neck. As my specimens have been compared with a good many skins from Heligoland, England,
Morocco, and North America, which are larger, and as I
have seen one skin from Southern Mexico that was entirely
like my bird, I believe that it is a tropical subspecies of the
Common Tern. It is also remarkable that this Tern is not
regularly found south of the Bahamas, and has not yet
been recorded further south than Jamaica. My specimens
measure:—Culmen 1·35 inch, wing 10·1, tail 5·4, tarsus 0·7.

[21]

The North-American birds were formerly called *Sterna wilsoni*, but they are absolutely identical with European ones.

37. STERNA ANTILLARUM, Less.

Common on Aruba and Bonaire in places where a sandy beach offers them good breeding-grounds. I believe they had laid their eggs on Aruba at the end of June, but I did not find any. At the end of July I found half- and full-grown young ones. In coloration the young bird is similar to that of *S. minuta*, and therefore requires no description.

Iris deep brown; bill yellow, with black tip; feet yellow.

38. STERNA DOUGALLI, Mont.

Cf. *Sterna dougalli gracilis*, Cory, Cat. W. Ind. B. pp. 82 & 135 (1892).

There was a large breeding-place of this Tern on the coral-reef on the coast of Aruba. The eggs are always three in number; they are deposited on the sand and on the green shore-plants which often cover the soil. The eggs vary to the same extent as those of *Sterna hirundo* and *S. paradisea* and other species of the family. The skins are in plumage identical with those from Mexico and other parts of the West Indies. Iris dark brown; bill blackish, basal half more or less orange-red; feet bright red. If there is no other character to distinguish *Sterna dougalli gracilis* but the colour of the bill, my birds might belong to that subspecies.

39. STERNA ANÆSTHETA, Scop.

A good many of this species were found breeding on the coral-reef off Cero Colorado on Aruba, at the same place where *S. dougalli* had its eggs. The eggs, however, were always laid in a corner under bushes, or under a stone or shell, and never placed so openly as those of *S. dougalli*. We found only one egg in each nest, and altogether not more than ten; they were all more or less set. When flying overhead the underpart of the wing and abdomen of this bird appeared beautifully tinged with greenish blue, while in the living *S. dougalli* the delicate peach-blossom colour was exceedingly pretty, but soon faded away after the birds were skinned. Iris deep brown; bill and feet black.

40. LARUS ATRICILLA, Linn.

Often seen on the coasts of Aruba, Curaçao, and Bonaire.

III. *Birds of Curaçao.*

Curaçao had been twice visited by collectors before my arrival, and two articles had been written on its birds. Ridgway (Proc. U. S. N. M. vii. pp. 173–177, 1884) enumerated 6 species. Berlepsch (J. f. O. 1892, p. 61) gave 19 species, the results of a collecting-tour made by Herr Peters, who appended to Berlepsch's admirable essay a list of 51 species supposed to occur on Curaçao. Of these 51 species, examples of 18 only were collected, and about 16 remained more or less doubtful or were founded on the erroneous information of the natives. Peters's list, however, contains some very useful field-notes, local names, and other information.

My collection contains examples of all the species that have been hitherto identified with certainty, except one [*].

I am greatly obliged for much help and kindness to Mijnheer Harry Barge, the Governor of the Dutch West Indies, to Mijnheer van der Linde Schotborgh, owner of the beautiful estate of Savonet, and to the chemist, Herr Ludwig, who takes an ardent interest in the natural history of Curaçao.

1. MIMUS GILVUS ROSTRATUS, Ridgw. Proc. U. S. N. M. 1884, p. 137; Berl. J. f. O. 1892, p. 74; Peters, *t. c.* p. 115.
Common. See above, p. 294.
Eggs were taken in June and July, but at the same time full-grown nestlings were found.

2. DENDRŒCA RUFOPILEATA, Ridgw. Proc. U. S. N. M. 1884, p. 173 (type from Curaçao); Berl. J. f. O. 1892, p. 76.
I have collected a series of twenty specimens of this bird

---

* *Ardea herodias*, Linn.

in Curaçao and Bonaire. It is equally common on both these islands, and is an inhabitant of open bushy places, as well as of mangroves and other trees. My series shows a very great variation. The adult males are bright yellow beneath; the breast, and sometimes the sides of the body, streaked with rufous; and the top of the head has a large patch of chestnut-brown. Sometimes the entire top of the head is covered with this colour, sometimes it forms a horse-shoe, sometimes it is developed only on the forehead and over the eyes. It seems that as the bird advances in age the chestnut on the head and the striations on the lower parts are more developed. Quite young birds have no streaks on the breast and no chestnut on the crown. The females, as a rule, have no chestnut on the head nor streaks beneath, but sometimes indications of the stripes and of the chestnut crown are visible, and in some specimens the top of the head is spotted with chestnut and the streaks on the breast are well developed, although not so strongly as in the adult males.

This species is very closely allied to *Dendrœca capitalis*, Lawr., from Barbados, but the chestnut on the crown is generally lighter, and the streaks on the breast are somewhat broader and not so well defined. Some specimens, however, run very close to those from Barbados.

This bird is very familiar, and known under the name of " *Para de misa*," which means " mass-bird," and often lives with great tameness in the vicinity of houses. Its song is a melodious warbling, soft and short, chiefly heard in the early morning. The nest is placed on the outer twigs of bushes, and is a tiny, very deep cup-shaped structure, composed of thin grasses interwoven with spider-webs, feathers, and hairs. I found some nests at the end of July, but did not get any eggs.

3. CERTHIOLA UROPYGIALIS (Berl.).

The nearest ally of this species is not *C. barbadensis*, as surmised by H. von Berlepsch (J. f. O. 1892, p. 77), but *C. newtoni*, from St. Croix, and *C. sancti-thomæ*. Berlepsch's

[24]

new species, however, can be distinguished by the large white
spot between the yellow breast and the slaty-black throat.
This white spot is extremely small and scarcely indicated in
C. newtoni, so small, in fact, that I have not found it
mentioned in any description (*cf.* Scl. Cat. B. xi. p. 43).
C. newtoni has also the uropygial band somewhat broader
and of a more yellowish-olive colour. The fresh unworn
specimens of *C. uropygialis* have distinct whitish edges to
the longest and some of the median wing-coverts.

It is easily distinguishable from *C. newtoni* by the much
darker throat, the white patch below the blackish throat,
the smaller white wing-speculum, and the colour of the
uropygial band. The plumage of the adult has been well
described by Berlepsch (*l. c.*). The young bird is grey
above, the uropygial band less developed, the crown similar
to or a little darker than the back ; beneath paler yellow,
throat pale grey mixed with yellow, superciliary stripe
yellow. In adult specimens the superciliary stripe is pure
white, as a rule, but many are found with the stripe more
or less tinted with yellow.

Berlepsch's Flower-pecker is extremely common on the
islands. It is called " *Barica-geel,*" i. e. Yellow-breast. In
Bonaire, at Mr. van den Brandhof's, they came into the
verandah to take milk and water and bread and fruits that
were offered them on a plate.

Its song is not loud nor attractive, being a metallic
warbling, frequently repeated.

The nest is a large ball of dry grasses and leaves, lined
with feathers, and with a lateral entrance. It is placed
mostly at the tips of branches at different heights from the
ground. One, from which I took eggs, was built in a flower-
basket hanging from the ceiling of Señor Ricardo's verandah
in Curaçao. The eggs were fresh, but the bird had left
them.

The eggs are four in number, of a whitish colour, more or
less thickly covered with darker and paler rufous spots and
patches. Their average size is 0.6×0.45 inch.

4. AMMODROMUS SAVANNARUM (Gm.).

Very rare on Curaçao, and only met with near Beekenburg, in a stony valley of grass and low bushes. Not previously recorded from Curaçao.

5. ZONOTRICHIA PILEATA (Bodd.); Berl. J. f. O. 1892, p. 82.

Local name " *Chonchorrongai*."

As a rule specimens of this bird from Curaçao are rather pale, but this character is not constant. It does not build closed nests, as suggested by Herr Peters (J. f. O. 1892, p. 115), but open ones, like other species of this genus. I found two eggs at the end of July. They are of a very pale-blue colour, regularly spotted with rufous. They measure 0·8 to 0·6 inch.

6. EUETHEIA SHARPEI, Hartert.

Euetheia bicolor, Berl. J. f. O. 1892, p. 81; Peters, t. c. p. 116.

Euetheia sharpei, Hart. Bull. B. O. C. vii. p. xxxvii.

Of all the birds that I collected on my West-Indian trip, those of the genus *Euetheia* (or *Phonipara*, as it is termed by Dr. Sharpe and others) are the most puzzling. After a careful comparison of all the materials at hand, I came to the following conclusions, and I believe that those ornithologists who have sufficient evidence to form an opinion will agree with me.

(1) Dr. Sharpe is correct in retaining as a separate subspecies *E. marchi*, notwithstanding that Mr. Cory has united all the West-Indian *Euetheiæ*. A fine additional series from San Domingo has arrived at the British Museum since the publication of the twelfth volume of the Catalogue. I think it will speak well for the distinctness of *E. marchi* when I say that, on a dark December day in London, I was able to pick out in a minute all the males of *E. marchi*, without mistake or hesitation, from the box containing *E. bicolor*, in which they had been provisionally placed. Besides the characters given by Dr. Sharpe in the ' Catalogue of Birds,' *E. marchi* evidently has the bill of a much lighter brown.

(2) The distribution of the two forms, *E. bicolor* and *E. marchi*, as given by Dr. Sharpe cannot be maintained. There is no doubt that the *Greater Antilles*, Jamaica, San Domingo (and Porto Rico?) are inhabited by *E. marchi*, but Dr. Sharpe was misled by insufficient materials into including St. Thomas in its range. I have shot several males on St. Thomas, which clearly show that this island is tenanted by *E. bicolor* proper, the same as the Bahaman form, which is the typical one. Dr. Sharpe now agrees with me that the bird from St. Thomas is *E. bicolor*, and not *E. marchi*; he further writes me that the only male from Santa Lucia in the British Museum is a badly made-up skin and difficult to determine, although it looks somewhat like *E. marchi*. The Barbadian bird, singularly enough, is, in Dr. Sharpe's opinion, *E. marchi*, while the other islands of the Lesser Antilles are inhabited by *E. bicolor*. This seems very curious, but the outlying island of Barbados differs geologically and zoologically in many respects from the Lesser Antilles (*cf.* Feilden, Ibis, 1889, p. 478); therefore it is not very remarkable that Barbados should have a different form of *Euetheia*, but possibly additional materials might show that it is not the same as *E. marchi*—unless it has been introduced, which is not likely, as it is so common on that island.

(3) A series of skins from Aruba, Curaçao, and Bonaire belongs to neither of these two forms. Berlepsch (J. f. O. 1892, p. 81) first pointed out the differences of this new form, but having received only one male he did not know whether these differences were constant or not. I have named it *E. sharpei*, in honour of Dr. Sharpe and his work on the Fringillidæ.

(4) The birds from Venezuela and Tobago are similar *inter se*, but differ slightly from the Bahaman form, to which they are most nearly allied. These therefore must stand as *E. omissa* (Jardine) (type *ex* Tobago).

(5) It might, on account of the close relationship of these forms, the not yet sufficiently defined distribution of them, and the possibility of the occurrence of intermediate forms,

be more convenient to treat them as subspecies; but I think that, as a rule, insular forms, which, on account of their isolation are not likely to interbreed or produce intermediate forms, should be regarded as species rather than as subspecies, even if the differences be small.

(6) Linnæus, in 1758, named Catesby's "Bahama Sparrow" *Fringilla zena,* but afterwards transferred this name to another member of his large group *Fringilla*—the *Spindalis zena* of the present epoch—and renamed the " Bahama Sparrow " *Fringilla bicolor.* According to the law of priority, both birds should bear the specific term " *bicolor,*" which could not cause any inconvenience, the one being a member of the Fringillidæ and the other of the Tanagridæ.

(7) The females of all these forms are similar, and to be distinguished only with the greatest difficulty.

The " *bicolor* "-group of the genus *Euetheia* consists therefore of the following species or subspecies :—

(1) EUETHEIA BICOLOR (Linn.).

♂. Forehead and crown dingy black, gradually shading off into the dusky olive of the back. Black of breast extending down along the abdomen. Bill blackish brown. Wing 1·9 to 2·05 inches.

Hab. Bahamas and most of the Lesser Antilles, accidentally in Southern Florida.

(2) E. MARCHII (Baird).

♂. Above similar to *E. bicolor,* but the black on the underparts much less extended, abdomen paler and without black. Bill paler brown. Wing 2·05 inches.

Hab. Jamaica, San Domingo (Barbados ?).

I have not seen specimens from Porto Rico, but they probably belong to this species.

(3) E. SHARPEI, Hartert.

♂. Beneath similar to *E. bicolor,* but the black above confined to the forehead and sides of the head; back and rump paler, a little more shaded with greyish; the black of the breast somewhat less deep and duller. Wing 2 to 2·15 inches.

Hab. Aruba, Curaçao, and Bonaire.

(4) E. omissa (Jardine).

Similar to *E. bicolor*, but the wing longer and the colour of the back and rump deeper and more of a greenish olive. Wing 2·15 to 2·2 inches. *Hab.* Venezuela, extending north to Tobago and parts of Colombia.

E. sharpei is very common on Curaçao. Its nest is a large ball of grass with a lateral entrance. All that I saw were placed in the prickly branches of the *Opuntia* and *Cereus*. I found from three to four eggs in the nest, which are whitish, with a very faint bluish hue, much speckled with rufous, and with a few deep brown spots. They measure from 0·65 × 0·46 to 0·7 × 0·5 inch.

7. Icterus xanthornus curaçaoensis (Ridgw.) ; Berl. J. f. O. 1892, p. 82.

Icterus curaçaoensis, Ridgw. Proc. U. S. N. M. 1884, p. 174.

Not rare.

8. Icterus icterus (Linn.).

Icterus vulgaris, subsp. ? Peters, J. f. O. 1892, p. 114.

Not numerous, but well known. Colours of Curaçao specimens very bright. Cory (Cat. W. Ind. B. p. 146) says the same of examples obtained in St. Thomas.

9. Hirundo erythrogastra (Bodd.) ; Peters, J. f. O. 1892, p. 117.

I saw a specimen that was skinned by Herr Ludwig, and which undoubtedly belonged to this species. I think it is only a visitor from the north, because Peters tells us that it was numerous at the end of August, while it was so rare during my visit that I only saw a few in the town and was not able to procure a specimen.

[29]

10. ELAINEA MARTINICA RIISII (Scl.).

Elainea riisii, Scl. P. Z. S. 1860, p. 314.

Elainea martinica, Scl. Cat. B. B. M. xiv. p. 141 ; Berl. J. f. O. 1892, p. 85 (Curaçao).

I procured three specimens of this bird on Mt. Christoffel, but did not see it anywhere else. Mrs. Hartert thinks she saw it on Bonaire, but no specimen was obtained. My skins are in better plumage than those collected by Herr Peters, but are also somewhat worn. They entirely agree with specimens from St. Thomas. Specimens from Guadeloupe and Dominica are slightly different, and it is advisable to recognize Sclater's *E. riisii* (afterwards, in the 'Catalogue of Birds,' united with *E. martinica* by the same author) as a subspecies.

This is another instance of Curaçao not having the continental form, but the West-Indian one, and also of a nearer relationship to the St.-Thomas avifauna than to that of the other Lesser Antilles.

11. MYIARCHUS BREVIPENNIS, Hartert, Bull. B. O. C. iii. p. xii.

Not very rare near Savonet and in other well-wooded places.

(Peters says (J. f. O. 1892, p. 118) that he saw a rather large species of Tyrant through his glasses. From his description it cannot be any of those that are as yet known from Curaçao.)

12. SUBLEGATUS GLABER, Scl. et Salv. ; Berl. J. f .O. 1892, p. 84 (Curaçao).

My specimens of this bird agree with the type from Caracas (Venezuela) in the British Museum. It occurs on all the three islands and is not rare, but is by no means common. The wings of my eight specimens measure 2·58 to 2·8 inches, mostly 2·6 and 2·65 inches.

This species can be distinguished without difficulty from *Sublegatus platyrhynchus* from Bahia, Brazil.

[30]

13. Tyrannus dominicensis (Gm.); Berl. J. f. O. 1892,
p. 86.

H. v. Berlepsch raised the question whether the birds of
Curaçao belong to the typical form of *Tyrannus dominicensis*
from the Greater Antilles or to the large-billed *T. rostratus*,
Scl., from the Lesser Antilles. I have collected a series
sufficient to show that they belong to the true *Tyrannus
dominicensis.*

This bird has the same name which it or its allies have
almost everywhere in the West Indies and South America,
" Pitirri" or " Pipirri." Its note, indeed, is exactly like its
name. It is common on Curaçao, especially near Savonet,
and may even be seen in the outskirts of Willemstad.

Sclater (Cat. B. xiv. p. 271) calls it *Tyrannus griseus*, but
I agree with Berlepsch and others that Gmelin's *Lanius
tyrannus β dominicensis*, given with habitat and distinguished
by description, should provide it with a name.

14. Chrysolampis mosquitus (Linn.); Berl. J. f. O. 1892,
p. 86.

Not rare on flowering trees and on the flowers of the aloe,
but less common than the next species.

15. Chlorostilbon caribæus, Lawr.; Berl. J. f. O. 1892,
p. 87.

The type of *C. caribæus* came from Curaçao. The specimens are indistinguishable from those from Venezuela
(generally called *C. atala*). The nest is a tiny structure
built on a small twig. I obtained two eggs from Herr
Ludwig. They are oval in shape, and in colour plain white
without gloss. They measure 0·4 × 0·29 inch, and weigh
17 milligramms.

16. Stenopsis cayennensis (Gm.); Berl. J. f. O. 1892,
p. 87.

Unfortunately I was not able to get an adult male, but
only a female and two young birds of this Nightjar. When
comparing my specimens with those in the British Museum,
I was unable to find any differences. The bird breeds on

Curaçao and Bonaire, but I did not see it on Aruba. It is not common, and is mostly found in dry and stony places with scanty vegetation.

17. CROTOPHAGA SULCIROSTRIS, Sw.

Not previously recorded from Curaçao.

I met with several of these birds near Savonet, and procured a few specimens. Its occurrence so far eastwards is very remarkable. I believe that it is resident on Curaçao. The stomach contained grasshoppers. Iris deep brown. Bill and feet black.

18. CONURUS PERTINAX (Linn.); Salvad. Cat. B. B. M. xx. p. 197; Berl. J. f. O. 1892, p. 88; Peters, J. f. O. 1892, p. 112.

Berlepsch (*l. c.*) gives his opinion that, on account of the peculiar fact that *Conurus pertinax* occurs on the two islands of Curaçao and St. Thomas, and apparently nowhere else, it is quite possible that its original home is Curaçao, where it seems to be more common than on St. Thomas. There are, however, other birds that occur on both these islands, so that I hesitate at present to accept this introduction-theory. On St. Thomas this lovely Parrakeet is restricted to the hills on the eastern side of the harbour (*cf.* A. & E. Newton, Ibis, 1859, p. 374), and at the present time it is said to be so rare that they are no longer caught for sale, while formerly they were brought to the steamers by the negroes. On Curaçao it is very numerous in the western parts of the island, but not so common, although by no means rare, in the eastern. The nests are mostly built in the large ants'-nests placed in trees, into which they dig holes.

The negroes take the young ones from the nests and keep them in cages. Large numbers are sold to the sailors.

The plumage of the adult bird is well described by Salvadori, but the descriptions of Finsch and many others are confusing, as they do not distinguish between *C. æruginosus* and *C. pertinax*. In the young of *C. pertinax* little of the beautiful orange-colour on the cheeks, which are brownish, is to be seen; the forehead is tinged with greenish and brownish, and the throat and upper breast are more tinged with greenish than in the adult bird. The orange-colour gradually spreads over the sides of the head from the lores and region under the eyes, and is assumed not by moult only, as some of my skins, as well as my observations on two living specimens that I brought home with me, clearly show. The bill in the adult bird is deep horn-brown, while in younger specimens the upper mandible is more or less pale and whitish. The sexes are quite similar.

19. BUTEO ALBICAUDATUS COLONUS, Berl. J. f. O. 1892, p. 91.

This name was proposed by Berlepsch for the Buzzard of Curaçao, of which he had a single young bird, probably in first plumage. Unfortunately I was unable to get an adult specimen of this bird, but I obtained a young male from Herr Ludwig and shot a young female on Bonaire. Both these specimens are quite similar to the one that is minutely described by H. v. Berlepsch, and the differences from the young of *Buteo albicaudatus* from other countries seem to be constant and well marked. I have seen the adult bird sailing over and around Mt. Christoffel on Curaçao, and twice on Aruba, but had no chance of shooting one. The iris in the young is brown, feet yellow, cere pale greenish. In the stomach I found the remains of small birds, and the natives on the islands say that it is very destructive to fowls. Native name "*Pata-lejo.*" The wings of my two specimens measure 15·1 and 15·2 inches, the tarsus 3·2.

The adults when flying high in the air looked like the continental species, but will probably turn out to be distinct.

20. TINNUNCULUS SPARVERIUS BREVIPENNIS, Berl. J. f. O. 1892, p. 91.

Not rare.

21. POLYBORUS CHERIWAY (Jacq.); Peters, J. f. O. 1892, p. 110.

Not rare.

22. STRIX FLAMMEA BARGEI, Hartert, Bull. B. O. C. iii. p. xiii; id. Ibis, 1893, p. 124.

Face white, a dark brown spot in front of the eye. Upper surface the same as in most of the European specimens. Beneath white, sparsely spotted with dark brown. Tail pale greyish isabelline, spotted with dark grey, and with four distinct blackish bars. Iris deep brown. Bill whitish flesh-colour, toes brown, claws deep brown. Total length about 12 inches, wing 9·7, tail 4·3, tarsus 2·2.

This insular form is entirely different from the Barn Owls of the West Indies, and also from the South-American form. In colour it is similar to many specimens from Europe, and also to some from the Pacific Islands, but in its small size it is only to be compared with the Galapagos species, which, however, is of an entirely different colour. I have only one specimen, which was caught for me by order of Mijnheer Harry Barge, Governor of the Dutch West Indies.

This Barn Owl is said to be not very rare in some of the rocky parts of Curaçao. Two specimens sent by Herr Ludwig agree with my own.

I do not know whether this Owl occurs on Aruba, but there appears to be another species of Owl on that island of only about half the size of it, and of an ashy colour. There is said to be another Owl on Curaçao, but of what kind I do not know.

23. COLUMBA GYMNOPHTHALMA, Temm.; Hartert, Bull. B. O. C. iii. p. xii; id. Ibis, 1893, p. 123.

Although there are examples of this Pigeon in the Museums of Paris and Leyden, and one stuffed specimen in the British Museum, neither its exact habitat nor anything of its life-history was known, and it has been several times confounded with *Columba picazuro* from Brazil, for instance by Herr v. Pelzeln (Orn. Bras. p. 274).

It was first described in 1811 by Temminck in Madame Knip's work, 'Les Pigeons,' on p. 48, and the figure (pl. xviii.) clearly represents this species, although the granulated bare orbital space and the thighs are wrongly coloured, and the belly and under tail-coverts are much too dark.

It was observed and noticed on Curaçao by E. Peters (J. f. O. 1892, p. 112) under the local name " *Ala blanco* " (not " *blanca,*" as Peters spells it), but specimens were not preserved, and the species was not identified.

In the living bird the bill is of a whitish flesh-colour, the iris deep orange-brown. Round the eye is a smooth bare ring of a bluish-grey colour; this ring is surrounded by a large granulated naked space of a dark reddish-brown colour, somewhat like an over-ripe strawberry. Feet raspberry-red. The lower surface of the bird is vinaceous grey, shading into ashy on the flanks and belly. Thighs and under tail-coverts greyish white. The broad white line along the wing has caused this bird to be named " *Ala blanco,*" or " White-wing," on these islands. I have five adult males and one young female. The latter has only an indication of the granules round the eye, and the beautiful scaly-looking white and blackish borders on the hind neck, and pale vinous and blackish borders to the feathers between the shoulders, are only slightly indicated, but this seems to be due rather to the immaturity of the specimen than to its sex.

The total length of the adult males is about 13 inches, the wing measures from 7·55 to 7·85, tail about 5·3, culmen 0·65, tarsus 1 to 1·1, middle toe 1·2 to 1·26.

This species is not rare on Aruba, and is common in some localities on Curaçao, where there are many trees; it is also common on Bonaire. According to Herr Peters it likewise occurs on the coast of Venezuela, where it is called " *Manglera,*" but this statement requires confirmation.

This beautiful Pigeon generally flies about in flocks, picking up its food from the ground as well as from the trees. Its note is a deep cooing, consisting of four sounds. I found a fresh-made nest on the 23rd July, but no eggs in it. I also shot young birds at this time, so I believe that they breed twice during the year. The nest is a loose structure, like that of *Columba palumbus,* and placed mostly in the mangroves, but sometimes in other trees. They are rather

2 A 2

shy birds, but can be shot in great numbers in very dry weather near the water. The Europeans and natives on the islands much appreciate its flesh as food, and it does well in captivity.

24. ZENAIDA VINACEO-RUFA, Ridgw. Proc. U. S. N. M. vii. p. 176; Berl. J. f. O. 1892, p. 95.

Extremely common on Aruba, also common on Curaçao, but most numerous on Bonaire. Peters (J. f. O. 1892, p. 113) mentions this species under three names—No. 13. "*Ala duro,*" No. 14. "*Blauw Duiff,*" and No. 15. "*Patruchi.*" All three names apply to *Zenaida vinaceo-rufa*. "*Ala duro*" is the most familiar name for the adult bird; "*Patruchi,*" a name that is by some of the islanders erroneously applied to *Eupsychortyx gouldi*, is less in use; and "*Blauw Duiff*" is the Dutch name, mostly given to the young bird, which many natives believe to be a distinct species. *Columba portoricensis* is sometimes called by the last name on Bonaire.

The young bird is more rusty above and beneath than the adult, most of the feathers have white edges and white lanceolate spots at the tips. The females are much darker in colour.

Wing of adult 5·2 to 5·5 inches. I found the nest—a flat and loose structure, like all Pigeons' nests—about 10 feet high in a dividivi-tree. The two eggs are ovate in shape (*cf.* Ridgw. Nomencl. Col. pl. xvi. fig. 1), and in colour plain white, with a faint gloss. The weights are 460 and 455 milligramms, and they measure 1·23 × 0·86 and 1·1 × 0·84 in.

25. LEPTOPTILA VERREAUXI (Bp.).

My honoured friend Count Tommaso Salvadori has kindly examined some of my skins of this bird, and refers them to *L. verreauxi*. The species is rather rare on Curaçao.

Wing 5·4 inches. Iris pale orange or yellowish brown, bill black, feet red.

I believe this to be Peters's No. 11 (J. f. O. p. 113), for the islanders call it "*Tortel Duiff,*" and there is no such thing as *Columba plumbea* on Curaçao.

[36]

26. COLUMBIGALLINA PASSERINA PERPALLIDA, Hartert.
See above, p. 304.
Extremely common.

27. EUPSYCHORTYX CRISTATUS (L.).
See above, p. 305.
Not rare. Often kept in confinement and sold for food.

28. ARDEA HERODIAS, Linn.; Ridgw. Proc. U.S. Nat.
Mus. 1884, p. 177.
Messrs. Benedict and Nye have obtained examples of this
species on Curaçao. I did not shoot it, but once saw a huge
Heron near Savonet at sunset, which I think belonged to
this species.

29. ARDEA CANDIDISSIMA, Gm.
White Herons are of irregular occurrence on Curaçao, and
as I shot *A. candidissima* on Aruba, I suppose that they
belong to this species.

30. BUTORIDES VIRESCENS (Linn.).
See above, p. 307.
I saw this bird several times on Curaçao.

31. TOTANUS MACULARIUS (Linn.).
Actitis macularia, Peters, J. f. O. 1892, p. 120.
I saw a few of these birds on the Schottegatt, but did not
shoot any.

32. HIMANTOPUS MEXICANUS (Müll.); Peters, J. f. O. 1892,
p. 121 ("teste Ludwig").
Flocks of old and young of this Stilt were seen in June on
the lagoon of Savonet. The immature birds were very young
indeed, and were probably bred on the island. The wing of
my adult male measures 8·8 inches, bill 2·65, tarsus 4·5;
female adult, wing 8·4, bill 2·66, tarsus 4. Iris bright red;
bill black; feet coral-red. In young birds the feet are paler,
the iris somewhat dull red, and the bill grey.

33. HÆMATOPUS PALLIATUS (Temm.); Peters, J. f. O. 1892,
p. 121.
Herr Ludwig has seen and shot examples of this species.

[37]

34. Pelecanus fuscus, Linn.
Occasionally seen on the coast.

35. Fregata aquila (Linn.).
Occasionally seen on the coast or sailing over the island.

36. Phalacrocorax brasilianus (Gm.).
Peters (J. f. O. 1892, p. 122) mentions that he saw a Cormorant which can hardly belong to any other species than this.

37. Sterna maxima, Bodd.
See above, p. 309.
A few of these Terns were seen on the coast.

38. Sterna hirundo, Linn.
A few Terns belonging to this species (or to *S. dougalli*, see above, p. 310) were seen on the Schottegatt and near Beekenburg.

39. Larus atricilla, Linn.
Seen on the harbour of Curaçao.

IV. *Birds of Bonaire.*

Bonaire, the most oceanic island of the three, is generally more wooded than the other two, although some parts of it are very bare.

Nothing has yet been published on the birds of Bonaire. Professor Martin, who stayed on the island for five days only, mentions that he saw *Columbigallina passerina* and a *Conurus* different from that of Aruba, as also from *C. pertinax*, and it will be seen that his surmise on this point was correct. We also know that Dr. A. A. Julien informed Mr. Lawrence that the *Chrysotis* of Aruba, which was described by the latter as *C. canifrons*, was common on Bonaire. It will be seen, however, that it is not the same, but an allied species.

I am obliged to several residents of Bonaire, above all to our kind host Mijnheer van den Brandhof, the Dutch Official of Bonaire, to Mijnheer Boyé, and Mijnheer Hachett, for much assistance during our visit to this island.

[38]

1. MARGAROPS FUSCATUS (Vieill.).

This typical West-Indian bird was common in the gardens near Fontein, on the north-east coast of Bonaire, but I saw it nowhere else. I have compared my skins with specimens from the Bahamas, Haiti, Porto Rico, and St. Thomas, and am not able to distinguish between them. My specimens are somewhat pale, but all are in, more or less, worn plumage, and there are quite similar ones from the Greater Antilles in the British Museum.

These "Tjutjubis" are peculiar birds, running and hopping quickly through the foliage, and sometimes making a great noise by chattering, warbling, and whistling together. They are, I believe, entirely fruit-eaters, for I did not find anything else in their stomachs, and are destructive to the fruits of the date-palms, of the *Carica papaya*, and other trees. They are so fond of the papaya-fruits that they used to come through the lattice of the window into the room when we had these fruits on the table and soon made away with them. The native name is "*Tjutjubi spagnol.*"

Iris yellowish white in adult birds, brown in the young ones; bill brownish horn-colour; feet light brown.

The occurrence of this species here is remarkable, especially as another subspecies, *Margarops fuscatus densirostris* (Vieill.), is found on the Lesser Antilles.

2. MIMUS GILVUS ROSTRATUS, Ridgw.
Common.

3. DENDRŒCA RUFOPILEATA, Ridgw.
Very common.

4. CERTHIOLA UROPYGIALIS (Berl.).
Extremely common.

5. AMMODROMUS SAVANNARUM (Gm.).
Common in grassy places on "Aruba-Estate," near Kralendijk, on Bonaire. It is called "*Raton de cero,*" or "*Para de cero.*" A series of skins of this species agree best with specimens from Jamaica, which are typical *A. savannarum*, and cannot be separated from them. The wings of the Aruban specimens measure 2·05 to 2·2 inches, tarsus 1·7.

[39]

The occurrence of this species here is very remarkable. Cory (Cat. W. Ind. B. p. 112, 1892) only gives Jamaica, Cuba, and Porto Rico as its habitats.

6. EUETHEIA SHARPEI, Hartert.
Very common. See above, p. 314.

7. ICTERUS XANTHORNUS CURAÇAOENSIS (Ridgw.).
Rather scarce on Bonaire.

8. MYIARCHUS BREVIPENNIS, Hartert.
Not rare. See above, p. 318.

9. SUBLEGATUS GLABER, Scl. et Salv.
Not rare. Called on Bonaire "*Para stranjero.*"

10. TYRANNUS DOMINICENSIS (Gm.).
Not rare, but perhaps less numerous than on Curaçao. Native name " *Pitirri.*"

11. CHRYSOLAMPIS MOSQUITUS (Linn.).
Common. In the middle of July most of the specimens, but not all, had passed through their moult.

12. CHLOROSTILBON CARIBÆUS, Lawr.
Common. Most of the specimens were still in moult, but a few very fine ones were shot.

13. STENOPSIS CAYENNENSIS (Gm.).
Rare. Native name "*Para de noche.*"   See above, p. 319.

14. CHRYSOTIS ROTHSCHILDI.   (Plate IX.)
Chrysotis rothschildi, Hartert, Bull. B. O. C. iii. p. xii; id. Ibis, 1893, p. 123.
The Amazon of Bonaire is allied to *Chrysotis ochroptera* from Venezuela and Aruba, but may be distinguished from it by the following characters :—
(1) Instead of the entire sides of the head being yellow (as in the adult *C. ochroptera*), only the anterior part of the crown, the space round the eyes, and the ear-coverts are yellow, and the green colour reaches up to, or nearly to, the lower mandible right and left of the chin.
(2) While in adult specimens of *C. ochroptera* the chin and entire throat are rich golden yellow, in *C. rothschildi*

no yellow feathers are to be seen on the throat, and only a few scanty feathers on the upper chin are of a pale yellow. In nearly all my specimens these feathers are of a more or less reddish-brown colour, but this, I believe, is due only to the juice of some fruit, as in one rather clean specimen they are pale yellow.

(3) While the whole of the band of the wing is yellow in *C. ochroptera,* and only a few scanty red feathers are sometimes to be seen next the body, the cubital edge in *C. rothschildi* is bright scarlet, more or less mixed with yellow outwards, but not to a great extent.

(4) The yellow shoulder-patch is very much smaller, and often quite restricted and mixed with red. The outer bend of the wing is not pure yellow, but yellowish green.

(5) The rump and abdomen show less or no blackish edges to the feathers, and the abdomen is less distinctly tinged with blue.

I may add that the skulls of *C. rothschildi* appear to be decidedly smaller, and that the bills are generally thinner and the wings somewhat shorter; but these are not very decisive characters for distinguishing this species, as they are not quite constant. In other respects *C. rothschildi* resembles *C. ochroptera.* Were it not for the red cubital edge and the less bluish tinge of the abdomen, *C. rothschildi* might be said to resemble a young stage of *C. ochroptera.*

The amount of scarlet at the base of the outer rectrices varies, and is sometimes spread over both webs of the second and third pairs of the outer rectrices.

In one quite adult female (No. 202 of my collection) some bluish feathers are visible on the forehead; they are perhaps the remains of the immature plumage, which I do not know.

Both Aruba and Bonaire must have received their Amazons from the continent. On Aruba, which is so close to the mainland, they have not become specialized, and very likely fresh immigrants might from time to time fly over to that island. Bonaire, however, is remote enough to produce a new insular form.

This Amazon is common near Fontein, on the N.E. coast of Bonaire, and is said to breed on rocks as well as in hollow trees. I am told that it also occurs on Mt. Brandaris, and that it straggles occasionally to different parts of the island.

These birds roost in the rocks of Fontein and fly out at day-break, returning to their roosting-places between 8 and 9 A.M. They leave again in search of food in the afternoon, and return just before sunset. They were easily shot when sitting on high trees or on the rocks, their harsh cries indicating their presence, although climbing, creeping, and shooting in the tropical heat and among those wild rocks is rather trying work. When feeding in the plains they appeared to be much more shy than when at home. Their food consists of fruits of the *Melocactus, Cereus, Morinda, Guava,* and other trees.

I shot nine specimens, but two were injured by shot and in moult, so that I brought home with me only seven skins. All are fully adult, and some more or less in moult. The sexes are alike.

The measurements are as follows :—

No.	Sex.	Wing.	Tail.	Culmen.	Height of upper mandible at base.
		in.	in.	in.	in.
188	♂ ad. sect.	8·5	5·4	1·4	0·65
189	♀ ad. sect.	7·85	5·1	?ᵃ	0·6
191	♂ sect.	8·5	5·1	1·35	0·59
198	♀ sect.	7·8	5·1	1·27	0·55
199	♂ sect.	8·0	4·9	1·26	0·6
200	♂ sect.	8·4	5·3	1·36	0·6
202	♀ ad. sect.	8·3	5·3	1·27	0·6

ᵃ Not measurable; tip wanting.

The eyes, bill, and feet are of the same colour as those of *C. ochroptera.* See above, p. 301.

15. CONURUS XANTHOGENIUS, Bp.

Bonaparte (Consp. i. p. 1) described *Conurus xanthogenius* from a single specimen, without locality, in the Leyden Museum. He gave Brazil as its habitat, but this, of course, was wrong. The careful description of the type specimen and notes on it in Finsch's work left me little doubt that *C. xanthogenius* was the same as my *Conurus* from Bonaire. To make sure I sent two of my specimens to my friend Büttikofer, who kindly compared them, and found them identical with the type of *C. xanthogenius*.

C. xanthogenius is similar to *C. pertinax*, except that in adult specimens the *entire top of the head is of a beautiful golden-yellow colour*, somewhat more orange on the forehead, while in *C. pertinax the forehead only is orange-yellow*. One specimen only having been known until lately, it was, in my opinion, quite reasonable to consider this form merely as an individual variety, as has been done by Finsch, Schlegel, Salvadori, and others. Since, however, all the adult specimens from Bonaire have the entire top of the head golden yellow, or at least strongly intermixed with golden yellow (all the new-coming feathers being of this colour), there can be no doubt that *C. xanthogenius* must stand as a distinct insular form. There is among the series of *C. pertinax* in the British and Leyden Museums, and among those collected by Herr Peters and myself on Curaçao, not one specimen with the top of the head yellow, although occasionally, but very rarely, a yellow feather appears there, chiefly in caged birds, as is common in Parrots, which are so much inclined to xanthochroism.

The young of *C. xanthogenius* are similar to the young of *C. pertinax*, but begin to show yellow feathers on the head at an early age. While the young examples of *C. pertinax* from Curaçao have the upper mandible always whitish, this part is brown (as in adult birds) in three immature specimens from Bonaire, but in one from the same locality it is more whitish.

It seems that the culmen in the Bonaire species is somewhat longer, as a rule. The measurements of twelve specimens from Bonaire are—total length about 10 inches, wing 5·5 to 5·8 (average 5·6), tail 5 to 5·5, culmen 0·9 to 1·06; while those of *C. pertinax* from Curaçao are—total length about 10 inches, wing 5·4 to 5·7, tail 4·7 to 5·6, culmen 0·83 to 0·95.

Having said so much about the Parrakeet of Bonaire, I must add that it is extremely common and numerous in almost every place in the island where the country is not quite bare. The screaming of these lovely birds is about the commonest noise that is heard in the bush on Bonaire, but they are often rather shy. The yellow head of the adult is so clearly visible that even a geological traveller like Professor Martin noticed it, and has distinctly said that the bird is of a species different from *C. pertinax*.

Whether ornithologists are inclined to call it a species or a subspecies matters little, but it is certainly different from the *Conurus* of Curaçao.

16. BUTEO ALBICAUDATUS COLONUS, Berl.
See above, p. 321. Rare on Bonaire.

17. TINNUNCULUS SPARVERIUS BREVIPENNIS, Berl.
See above, p. 321. Very rare on Bonaire.

18. POLYBORUS CHERIWAY (Jacq.).
Occurs everywhere, but not in large numbers. These birds are often killed because they are supposed to destroy the chickens. Besides the preceding species, I conclude from the reports of the inhabitants that several other birds of prey, and among these *Falco peregrinus* and *Pandion haliaëtus*, visit this island, as well as Aruba and Curaçao, in the winter.

There appear to be no Owls in Bonaire.

19. COLUMBA GYMNOPHTHALMA, Temm.
Very common.

20. COLUMBA PORTORICENSIS, Temm.
Columba portoricensis, Temm. in Knip's 'Les Pigeons,' pl. xv. p. 41.
Columba corensis, auctorum.

[44]

My friend Hans von Berlepsch has called my attention to
the original description of Jacquin's *Columba corensis* (Beytr.
z. Geschichte d. Vögel, 1784, p. 31). The author there says,
"*Columba* (*corensis*) *cauda æquali, orbitis denudatis atro-
punctatis, corpore grisco.*"

"Near Koro, in Venezuela, occurs a fine Pigeon, which
agrees in size with the common domesticated Pigeon. It
is entirely of a beautiful grey colour, and the feathers of the
hind neck are scale-like, which, although of the same colour
as the others, appear different in different lights. The red
eyes stand in a bare space, which is beset with black spots.
The feet are red. The Indians take the young from their
nests, feed them up, and eat them." (*Translated from the
German.*) Gmelin's diagnosis is merely based on Jacquin's
description, and I quite agree with Berlepsch that the
description is so uncertain—the more so when considering
that the West-Indian *Columba corensis* of recent authors
has not yet been found on the continent—that the name
of Temminck, who gives a good figure and description of it,
should stand for this species. I am very glad to learn that
Count Salvadori agrees with us in this conclusion.

Examples from Bonaire are absolutely identical with specimens from Cuba.

I met with this Pigeon only among the rocks on Bonaire,
where it is fairly common near Fontein.

Its note is a very loud and strong cooing, consisting of
three sounds, somewhat like *coo-róo-coo,* and repeated very
frequently.

I did not see this Pigeon on the ground, and it appears to
get most of its food from the trees.

The bill is of a dark blood-red colour, horn-white at the
tip. The iris consists of two rings, the outer one crimson,
the inner one yellow. The naked papillose space round the
eye is *yellow*, not red.

There is said to be a Pigeon on the Christoffel in Curaçao,
of which neither Herr Peters nor I have been able to get
specimens. It is called "*Paloma preto,*" which means
"Black Pigeon." A native of Curaçao told me it was the

same as the Rock Pigeon *("Paloma di barranca"* or *"Paloma blauw")* of Bonaire, and Herr Ludwig, of Curaçao, wrote me to the same effect.

21. ZENAIDA VINACEO-RUFA, Ridgw.
Very common. Large numbers are shot in dry weather at the water-tanks, and they make an excellent dish, as does also the *Columbigallina.*

22. LEPTOPTILA VERREAUXI, Bp.
More common than on Aruba and Curaçao.
They are tamer than the other Pigeons and very easily shot. The natives call it *" Toewiri"* on Bonaire, and sometimes *" Pecho blanco."*

23. COLUMBIGALLINA PASSERINA PERPALLIDA, Hartert.
Extremely common everywhere. In dry weather they assemble at the edge of the water-tanks in such numbers that sometimes 50 or 60, or even more, can be killed with one shot. The islanders use specially loaded cartridges, containing little powder and much of the smallest shot, for these slaughters.

24. ARDEA TRICOLOR, Müll.
Seen once or twice on Bonaire.

25. ARDEA CANDIDISSIMA, Gm.
Not rare near the "salt-pans" in the south of the island.

26. BUTORIDES VIRESCENS (Linn.).
Seen once or twice in the south of Bonaire.

27. HIMANTOPUS MEXICANUS (Müll.).
I saw a flock of these birds in the south of the island.

28. TOTANUS MELANOLEUCUS.
On the 21st July, when on the island of Bonaire, three of these birds passed overhead. I was able to fire only one shot, which brought down one of them. This is an adult male, and agrees perfectly with specimens from other localities, except that its wings are shorter, measuring only 6·95 inches, and the tarsi only 2·2. The bill is of the same length as

that of other examples, measuring 2·2 inches. If this should prove to be a resident bird, it is not unlikely to be a short-winged insular form.

Iris deep brown ; bill black, dark greyish horn-colour at the base ; legs yellowish.

29. TRINGA MINUTILLA, Vieill.

A single male of this bird was noticed and shot on the 23rd July at Laguna, on Bonaire. It agrees with specimens from other localities.

30. ÆGIALITES COLLARIS (Vieill.).

Small flocks of this bird were seen on Bonaire, and two young specimens in moult were procured. It seems to me that the wings and bills are rather short, but examination of a series of adult specimens would be necessary to guarantee the constancy of these characters. The culmens measure 0·53 and 0·57 inch.

31. ÆGIALITIS RUFINUCHA, Ridgw.

Rather common at Laguna and at the "salt-pans," where they undoubtedly breed. Bonaire specimens are like those from Aruba.

See above, p. 307.

32. PHŒNICOPTERUS sp. inc.

A great number of Flamingoes breed on Bonaire. They are locally called "*Choyogo.*" On the 12th June I went to the "salt-pans," where I saw several hundreds of Flamingoes standing in the middle of the vast shallow water-basin on their nests. Unfortunately I had no rifle with me, and the locality not producing a single bush nor anything to hide anyone approaching, it was impossible to get within gunshot-distance. The aspect of hundreds of these wonderful birds was even more picturesque than that of the Indian Flamingo. In spite of the assurances of the men, who told me there were no eggs, I walked, along with my guide, knee-deep through water (which was, in fact, like a solution of salt and saltpetre) to the nests. The travelling was very unpleasant, not to say dangerous. The water was deep in places and the bottom very rough, consisting of very sharp corals, and often of a deceitful crust of salt or saltpetre, under which the water was black and very deep. It required much care to avoid these bad places, and it took us, I think, nearly an hour to reach the nests. Our shoes being cut by the corals, our feet began to bleed, and the salt water caused an unpleasant tingling in the little wounds. The nests themselves were flat plateaus, standing out of the water from three to six inches, the water round them being apparently very shallow, but it was often the fatal crust that caused this appearance, not the proper bottom. Many of the nests were close together and sometimes connected by dry ground. They were quite hard, so that one could stand on them, and almost the only way of getting along was to jump from one nest to the other. The nests consisted of clay, hardened by the sun and penetrated and overcrusted with salt, and of pieces of coral, with a distinct concavity in the centre. On some of the nests we found freshly-broken eggs of some species of Tern, and lying in the water I found two eggs of the Flamingo, which turned out to be quite fresh and eatable, although they must have been in the water for some time. After the breeding-place of the " *Chogogo* " had been thus disturbed, these shy birds left the spot and flew to the other side of the island. I am told that they change their breeding-places very often. The two eggs measure 3·35 and 3·45 inches by 2·13 and 2·16.

Except leaving them of a colour like that of a boiled lobster, this pleasant trip had no evil result on my legs, but my guide, the faithful policeman of Mijnheer van den Brandhof, lost the entire skin of his, and could not go out for some days afterwards.

33. Pelecanus fuscus, Linn.
I saw flocks of this Pelican at sea close to the shore.

34. Fregata aquila (Linn.).
I did not see the Frigate-bird on the island myself, but I am assured that it not rarely occurs there.

[48]

35. STERNA MAXIMA, Bodd.

See above, p. 326.

36. STERNA HIRUNDO, Linn.

I have already mentioned that I found the broken eggs of some Terns on the nests of the Flamingoes; but I regret that I was so much occupied and excited by the Flamingoes and their breeding-place that I did not pay sufficient attention to the Terns to say with certainty whether *S. dougalli* is found here as well as *S. hirundo*. Two Terns that came near to me and were shot were of the latter species, and therefore I am quite sure that they breed here and that the broken eggs belonged to them.

37. STERNA ANTILLARUM, Less.

These Terns were common, and had both nearly and quite full-grown young ones.

38. LARUS ATRICILLA, Linn.

This Gull was seen several times on the coast.

V. *General Conclusions.*

(1) The three islands of Aruba, Curaçao, and Bonaire have received the greater number of their birds from the South-American continent, but some also from the West Indies, for there are many pure West-Indian forms amongst them besides the continental ones.

(2) There are striking affinities between the avifauna of these islands and that of the islands of St. Thomas and St. Croix (Virgin Islands), but no similarity to that of the Windward Islands; for example, *Conurus pertinax, Elainea martinica riisii, Icterus icterus*, Tyrannus dominicensis†,* and *Margarops fuscatus* occur in both localities. Moreover, we have in this avifauna *Certhiola uropygialis* (of which the nearest allies are found on St. Croix and St. Thomas), *Ammo-*

---

* It has been suggested that *Icterus icterus* has been introduced into St. Thomas, but this seems to be doubtful.

† *T. dominicensis* is replaced by *T. rostratus*, Scl., on most of the Lesser Antilles.

dromus savannarum*, and *Eupsychortyx cristatus*†. These facts are very interesting and should be studied more thoroughly : they seem to point to the theory that the Virgin Islands and the islands of Bonaire and Curaçao ‡ were formerly connected in some way, or that they are of the same geological age, and not of the same age as the Windward Islands.   Perhaps there was once a line of islands (similar to that of the Lesser Antilles) reaching from St. Thomas through "Los Aves," or the Bird Island, by way of Blanca, Orchilla, Grand Cay, Los Roques, and the second group called "Los Aves," to Bonaire and Curaçao.

(3) The avifaunas of the three islands are generally very similar, but some interesting differences are obvious.   Bonaire has most species of West-Indian origin, while Aruba has most continental forms, as would be expected from their situation.

(4) The facts brought to light through my little collection should induce ornithologists to explore the other small islands on the Venezuelan coast, such as Grand Cay, Orchilla, Blanca, and Margarita.

---

* *Ammodromus savannarum* is not found in the Lesser Antilles, but occurs in Porto Rico, very near to St. Thomas.   See Cory, Cat. W. Ind. B. p. 112.

† It is said that, although *E. sonninii* occurs on St. Thomas, it has been introduced from Venezuela (*cf.* Cassin, Proc. Ac. Nat. Sci. Philad. 1860, p. 378; Newton, Ibis, 1860, p. 308; and Berl. J. f. O. 1892, p. 100).   If this is correct, no weight should be attached to the occurrence of a different species (*E. cristatus*) in the Curaçao group.

‡ To Aruba these species may have been brought by the trade-wind from the other two islands.

www.ingramcontent.com/pod-product-compliance
Lightning Source LLC
Chambersburg PA
CBHW031815090426
42739CB00008B/1287